Printed in the USA
CPSIA information can be obtained
at www.ICGtesting.com
LVHW081644151024
793887LV00015B/710

9 781694 705303